THE NAKAMURA LETTERS

FRANKIE BOW

The Nakamura Letters

Published by Hawaiian Heritage Press

ACKNOWLEDGMENTS

Thanks to the Hawai'i Department of Land and Natural Resources Division of State Parks, without whom the inspiration for this story would not exist.

CHAPTER ONE

Emma Kano`opomaika`i Nakamura <emmakn@mahina.edu>

to Molly

Molly,

So I just got here and haven't even unpacked yet, but I'm writing you first thing just like I promised.

So here I am in my exotic sabbatical location a 2 hour drive from Mahina. We'll see how long it takes this email to get to you. I'm writing it at 2:54pm on Monday. The only internet access I have here is through my cell phone. The signal is junk, but least there's a landline. At least I think there is, I'll have to try the phone later to make sure it works.

The fridge is old and has rust on it but the temp is fine and there's enough room for my samples.

The water's catchment, they got one big old concrete tank out back. There's no UV filter, so I gotta boil all my water just to be safe.

No idea why someone decided to build a house way up here a hundred years ago or whenever, or how the park ended up owning it, but it works out perfect for my research.

So you getting used to being a mom yet? Say howzit to Donnie for me. (Just kidding I can't imagine you saying howzit. Say Good Morrow, Dear Husband or whatever it is you call him.)

Hope you appreciate your fast internet and clean county water. And your warm temperatures too, did I mention I'm freezing my okole off up here? I'll call you tomorrow and you can try get my landline number from your caller ID. I can't find the phone number anywhere. If it was ever on the phone it wore off years ago. Did I mention this place is old?

PS did Kayla start yet? You'll like her, Molly. Once you get some regular help with the baby you'll start to feel human again.

Emma Nakamura, PhD

Professor of Biology

Mahina State University

mahina.edu

Some of the most fun people I know are scientists.

Mae C. Jemison (b. 1956)

CHAPTER TWO

Emma Kano`opomaika`i Nakamura <emmakn@mahina.edu>

to Molly

Eh Molly,

No worries about being cranky. I guess I'd be cranky too if I had a bunch of guys revving up their chainsaws outside my window all day while I was "housebound, leaking from every orifice, hooked up to wheezing pumps like a dairy cow, and at the round-the-clock beck and call of an insatiable eight-pound nipple-crushing lamprey."

I actually wrote down what you said for the next time Yoshi brings up having kids.

Good thing you guys are finally cutting back that hedge though. I can't believe how fast that thing grew, it was taller than your house last time I was over there. That's the thing about gardening in Hawaii. You bring over some well-behaved plant from the mainland and it gets here and suddenly it's out of control and climbing all over everything. Kind of like our exchange students.

You'll be glad you got the hedge cut down. It was making your house kind of dark, and plus there'll be less places for coqui frogs to hide.

That's one good thing about staying way up here, there's a lot less coquis. I heard one or two last night, but it wasn't even close to being as bad as it is down at sea level.

I don't think you need to worry about Kayla telling everyone on campus about your private life. First of all, your private life isn't that interesting. Especially not for the first six weeks, if you know what I mean. Second, Kayla's doing an official practicum through the school for her early childhood ed program, so if she does anything unprofessional like blab all your secrets (if you had any) she'd get kicked out. Third, and most importantly, you live in Mahina, Molly, what secrets do you think you have left?

I still don't understand how come you don't get some kind of leave or something for the baby. That's so messed up. At least they're letting you teach your classes online so no danger of you showing up to class and accidentally spraying the front row with breast milk.

There was a freak lightning storm today so I was stuck inside. It gave me a chance to look around the house some more. Molly, I don't think anyone's been here for like decades. I found some Yardley of London (?) lipstick in one of the bathrooms. It looked super retro and smelled like old crayons. This whole place is creepy. You'd love it.

I can't believe how long it took for my last email to get to you. I guess my emails cue up and don't get sent out till an angel flies overhead at just the right altitude or something. Well at least you're getting them.

How was Kayla's first day? She's good, yeah?

Emma Nakamura, PhD

Professor of Biology

Mahina State University

mahina.edu

Science is the great antidote to the poison of enthusiasm and superstition.

Adam Smith (1723-90)

CHAPTER THREE

Emma Kano`opomaika`i Nakamura <emmakn@mahina.edu>

to Molly

Honestly Molly I don't think you're doing this mother's helper thing right. If you feel like you have to stay and watch Kayla whenever she's watching the baby, why have her there at all? And no I don't believe she's after Donnie, that's your hormones talking. Come on Molly, she was one of my best students, I wouldn't have told you to hire some sleaze.

Seriously, leave the baby with her and Donnie for a couple days and come up to visit me here.

Oh and when you come up, bring pizza, cause no one delivers all the way up here. I already checked.

OK, guess you'll get this when it QUEUES up, not cues up, is that better? (I still think CUES up is right. Like, it's your turn, that's your cue. But yeah, I'll give you the less vs. fewer thing. Fewer coqui frogs, not less coqui frogs. Happy?)

Emma Nakamura, PhD

Professor of Biology

Mahina State University

mahina.edu

Science and everyday life cannot and should not be separated.

Rosalind Franklin (1920-1958)

CHAPTER FOUR

Emma Kano`opomaika`i Nakamura <em-
makn@mahina.edu>

to Molly

So I forgot to tell you, Yoshi actually got that residential fellowship he was applying for. I know, I couldn't believe it either. He's gonna be on the mainland for three months, freezing his okole off in the woods working on his Vision or whatever. I told him he could come up here and stay with me if he wanted, it's plenty isolated, but then I found out they're paying him so I was like, bon voyage honey, see you when you get back.

It's weird, usually people bug me (not you so much, only sometimes) but it's so isolated up here, I feel like I'm about to go nuts. At least there's a park ranger. She came by today to say hi (probably check on me and make sure I'm legit). She seems okay. She told me her name but I forgot it right away. For some reason she thought it was important to tell me that the owner killed themselves here in the house, before it became part of the park. And then she goes, yeah, people are upset by that but why be upset at one suicide, what about all the other people who died in this house over the years, probably in the exact bed you're sleeping in.

Guess I'll sleep well tonight…

P.S. and yes, it is unreasonable and paranoid of you to forbid Kayla from wearing shorts at your house. It's freaking hot in Mahina, Molly.

Emma Nakamura, PhD

Professor of Biology

Mahina State University

mahina.edu

There is no adequate defense, except stupidity, against the impact of a new idea.

Percy Williams Bridgman (1882-1961)

CHAPTER FIVE

Emma Kano`opomaika`i Nakamura <emmakn@mahina.edu>

to Molly

The weather cleared up today, so I was able to get some samples. I'm not going to go into a bunch of detail cause I know you wouldn't understand the technical details, but I'm feeling pretty good about it. Oh yeah, and there's a little church down the road, closer to the highway, although I don't know who goes there, there's no other houses or anything around here. Oh, and good thing I brought the generator cause the power went out practically as soon as I'd gotten everything in the freezer. I'm typing this in the total dark in the middle of the living room, with my ear plugs in cause the generator's so loud.

Oh yeah, remember I got that billion-lumen whatever flashlight to bring up? Good thing I did. It's my only light now, besides my phone, so I'm walking down the hall to the bathroom hoping not to get the attention of any ghosts like I'm in The Sixth Sense or something. By the way, not like you asked for my advice, but unless you're going for sainthood (that's a thing Catholics do, right?) I don't think you should have to keep teaching your classes while you're on maternity leave. If your department doesn't have the money to run the classes your students need, that's the administration's problem, not yours. If you keep doing unpaid work for them, they'll just keep expecting it.

Of course I'm one to talk, look where I am. For sure no one's paying me extra to spend my sabbatical up here on the set of Friday the 13th:The Wilderness Years.

I was wondering whether I should tell you this or not so here goes: Last night when it was raining I thought I heard someone crying outside.

I'm sure it was a feral cat or something, but it kind of freaked me out. Just goes to show how your mind can go all weird on you when you're isolated.

OK, time for me to go to bed in complete darkness, and try not to think about all the people who died in this house. I'll write again as soon as I can cause I don't want you to go crazy bored at home and end up sticking your head in the oven. I don't need you haunting me on top of everything else I have to deal with.

Emma Nakamura, PhD

Professor of Biology

Mahina State University

mahina.edu

A ship in port is safe, but that's not what ships are built for.

Grace Hopper (1906-1992)

CHAPTER SIX

Emma Kano`opomaika`i Nakamura <em-makn@mahina.edu>

to Molly

You'll never believe it, I had another visitor today. Wow, two whole people in the space of three days, it's practically Grand Central Station up here. So I was out all morning. I came back kind of tired but I put away my samples, fixed some lunch, cracked a beer, and I just settled in to relax and catch up on my reading, and next thing I know there's someone's knocking on the door. So I thought maybe it was that park ranger again (I still can't remember her name) reminding me to expect to see blood running out of my faucets or furniture flying around the room or whatever, but no, it was this little old haole lady in a big straw hat. At first I assumed she was a tourist who got lost. But no, she came to this house on purpose. The Hodges House, she called it. I didn't know it actually had a name, but according to her it did. She was so excited I was staying here, she told me she was sure the house was lonely with no one staying there for such a long time. So that was creepy. (At least she didn't say the house was hungry, I guess that would have been worse. Cause by this point I could totally imagine this place chewing me up and spitting out my bones.)

So I invited her in and asked her how she happened to be in the neighborhood and she told me she'd been visiting the graveyard (??). I was all, what graveyard? but I guess there's one in back of that little church. I told you about that, right? Betty (that's her name) said her great-great-grandfather or someone was the first minister at that church so she has a bunch of family buried there.

Seriously, Molly, you should come up here and see this place. Bring Pat too, he'd love to do a story on it for Island Confidential I bet.

And bring pizza. And beer! I thought I packed plenty beer but I'm already running low.

Emma Nakamura, PhD

Professor of Biology

Mahina State University

mahina.edu

We must have perseverance and above all confidence in ourselves. We must believe that we are gifted for something and that this thing must be attained.

Marie Curie (1867-1934)

CHAPTER SEVEN

Emma Kano`opomaika`i Nakamura <em-makn@mahina.edu>

to Molly

Oh, you think Donnie would be “worried” about you and Pat making that long drive in your “delicate” condition, huh? Riiight. I gotta talk to that boy. He knows there’s no chance of you and Pat running off together, doesn’t he? Maybe he’s still holding a grudge about Pat giving you that copy of The Yellow Wallpaper for your wedding gift. Donnie shouldn’t take it personally though. Pat would’ve given you it no matter who you ended up marrying.

Hey, at least your husband likes having you around. Yoshi wasn't even considering coming up here with me cause he says he needs a reliable internet connection. I think the real reason was he doesn't like the idea of just him and me in the middle of nowhere with nothing to do but hang out with each other. Well, he's off in Indiana or someplace now, doing Art and getting a stipend and I guess atoning for his MBA years.

Tho to be fair to Yoshi (which I always am) he didn't want me to come up here either. He said he didn't think it was "wholesome" whatever that means. Probably he thought it was barbaric to stay somewhere with no cell phone service.

Wait a minute, did you actually ask Donnie if he minds if you come up for a visit? Or are you just assuming you know what he'd say? (You know what happens when you "assume," right?) Ask him, Molly! Bet he says yes.

By the way there is a cat that hangs around here. Brown with stripes, cause I guess a black cat would be too on-the-nose. Anyway that's probably what I heard that night that sounded like someone crying.

Emma Nakamura, PhD

Professor of Biology

Mahina State University

mahina.edu

Above all, don't fear difficult moments. The best comes from them.

Rita Levi-Montalcini (1909-2012)

CHAPTER EIGHT

Emma Kano`opomaika`i Nakamura <em-makn@mahina.edu>

to Molly

Oh, Donnie said it was fine with him for you to come? Told you he'd be OK with it. Yeah, I get the whole thing about you having to pump milk, in case you forgot I am a BIOLOGY PROFESSOR.

So pack your dairy equipment and get up here with my pizza and beer. Tell Pat to get his skinny tochas up here too. I miss you guys. There, I said it.

. . .

Emma Nakamura, PhD

Professor of Biology

Mahina State University

mahina.edu

Life need not be easy, provided only that it is not empty.

Lise Meitner (1878-1968)

CHAPTER NINE

Emma Kano`opomaika`i Nakamura <em-makn@mahina.edu>

to Molly

Eh Molly, you better get up here quick before I totally lose it. Stupid landline rang last night. One ring, and then nothing. And the technology's so old I don't think I can put any kind of caller ID on it. Something else, I managed to lose my reading glasses and they turned up later in the weird little bathroom I never use. Ever since then I've been imagining that things aren't exactly where I left them, although why would a burglar break in just to move my things and not steal anything?

At least Nettie seems happy with the service I'm providing. She left me a tip this morning. I tried attaching a picture but no can. Stupid data. So I'll just tell you, it's a dead rat.

(Nettie is the cat. If she was going to keep hanging around I thought I might as well call her something, so I decided to name her after Nettie Stevens.)

So yeah, after all this time I actually have a pet, kind of. I know, you're thinking where's Emma and what have you done with her, right? When the cat starts talking to me that's when you can call the guys with the butterfly nets.

And speaking of the guys with the butterfly nets, this might sound crazy, but with the phone calls and my stuff getting misplaced, I'm going to try an experiment: Sprinkle baby powder on the floor tonight and see if there are footprints tomorrow morning. I got the idea from Scooby-Doo.

Emma Nakamura, PhD

Professor of Biology

Mahina State University

mahina.edu

There are in fact two things, science and opinion; the former begets knowledge, the latter ignorance.

Hippocrates

CHAPTER TEN

Emma Kano`opomaika`i Nakamura <emmakn@mahina.edu>

to Molly

so it's morning and the results of the baby powder are…drum roll please…FOOTPRINTS!

Whose footprints you ask?

Stop drum roll, start sad trombone: MY Footprints. That's right. I guess I went downstairs in the middle of the night to check on something and forgot about it.

Molly, get your milk-soaked carcass up here before I totally lose it.

P.S. HOW had you not heard of Nettie Stevens? The discoverer of sex chromosomes? What are they teaching English majors these days?

Emma Nakamura, PhD

Professor of Biology

Mahina State University

mahina.edu

My mother always taught us that if people don't agree with you, the important thing is to listen to them. But if you've listened to them carefully and you still think that you're right, then you must have the courage of your convictions.

Jane Goodall (b. 1934)

CHAPTER ELEVEN

Emma Kano`opomaika`i Nakamura <emmakn@mahina.edu>

to Molly

So my new friend Betty Brewster came by again today. Remember, the little old haole lady with the giant straw hat? She practically invited herself in and acted all buddy-buddy with Nettie, who she called Balthazar (!?) (like any self-respecting cat would put up with being named Balthazar!)

Anyway all I had to give her was protein bars and instant coffee (yes, instant coffee. I know you're on your fainting couch right now). I gotta tell you, the kitchen's nothing to write home about (except I just realized that's exactly what I'm doing right now: Irony!) a 1950s stove, old cabinets, and a wooden floor that gives you splinters when you walk barefoot.

Even boiling water is kind of a pain because it turns out one of the burners goes on maximum heat when you turn it on no matter where you set the dial, one of them doesn't heat up at all, and the other two kind of turn on and off randomly.

I'm getting pretty sick of this place already, to be honest, but I'm nowhere near collecting all the samples I need. I want to just get it over with and then walk away. I should probably do a shopping run pretty soon, but it's like a forty minute drive down to the Hashizaki Store, and another 20 to the closest town. I feel like if I'm gonna drive two hours I should just drive back to Mahina.

Hey, you know what would be perfect? HALLOWEEN PARTY! This house is spooky enough for anyone. And you know there won't be any trick or treaters coming by so we can just hang out and drink all night, just as nature intended.

Update:

Right after I went to bed the phone rang. I ran downstairs answered it and I think I heard breathing, then it went dead. Say what you want about conserving energy, I'm sleeping with ALL the lights on tonight. WHEN ARE YOU COMING UP TO VISIT

Emma Nakamura, PhD

Professor of Biology

Mahina State University

mahina.edu

If you know you are on the right track, if you have this inner knowledge, then nobody can turn you off... no matter what they say.

Barbara McClintock (1902-1992)

CHAPTER TWELVE

Emma Kano`opomaika`i Nakamura <em-makn@mahina.edu>

to Molly

Molly, you will not believe this. So you know I was complaining about the long drive I'd have to make to get more beer and food and stuff, but I figured I might as well just get it over with. So I drove down the access road toward the highway and right past the church there was a barricade with the guys in orange vests, the whole thing, and know why? Turns out there was an accident with a propane truck down on the highway

RIGHT at the corner of the access road. MY access road.

You remember last time a propane tanker caught on fire here? In case you don't it burned for DAYS. Now I'm totally stuck. I can't get down, you can't come up. There's no other way out.

At least the power's on, and I have enough gas for the generator, for now.

Emma Nakamura, PhD

Professor of Biology

Mahina State University

mahina.edu

I hadn't been aware that there were doors closed to me until I started knocking on them.

Gertrude B. Elion (1918-1999)

CHAPTER THIRTEEN

Emma Kano`opomaika`i Nakamura <em-makn@mahina.edu>

to Molly

Eh Molly,

So here I am stuck in this old haunted house, miles from civilization, and who should stop by but my old friend Betty Brewster! Like, I'm stranded here by a flaming propane truck and somehow Betty manages to find her way here, how does that happen? And yes, it's her in person and no I'm not hallucinating. So first thing I asked was, where'd you come from and how'd you get up here? Did they put the fire out already? The answer was no, because of course it wasn't, the fire's inside the tank and they have to let it burn itself out. So she knew what was going on. Then she said, get this, she was camping up here. Come on, she just HAPPENS to be camping close to the house? Call me paranoid, but I think she's keeping an eye on the house now that I'm here.

And now that I think of it, I even think she might be the one who's ringing my phone.

So anyway I asked her if there was anything she needed (I'm sure it wasn't my instant coffee that lured her back). She said, well, you know it's Sunday. I actually didn't know off hand (I lost track of the days honestly) but I just nodded like I was keeping up. She said because the church is on the access road and blocked off from the main highway, the pastor and the audience can't get up there so she wanted me to go down there with her.

It didn't really make sense to me but it's not like my social calendar was exactly bursting with thrilling activities. I told her sure, I'll go down there with you. In fact I'm leaving in a few minutes. So if this is the last email you get from me, you know where to tell the authorities to look.

OK she's done freshening up talk 2U later if I'm not murdered.

Emma Nakamura, PhD

Professor of Biology

Mahina State University

mahina.edu

There is no law except the law that there is no law.

John Archibald Wheeler (1911-2008)

CHAPTER FOURTEEN

Emma Kano`opomaika`i Nakamura <emmakn@mahina.edu>

to Molly

OK first thing is you'll be glad to know I'm still alive.

OR AM I…?

I guess I probably wouldn't be able to type this if I was a ghost so yeah.

Second thing, how am I supposed to know it's called a "congregation" and not an "audience?" Plus I'm not sure you're right about that. What about when people have an audience with the Pope? I've seen it on TV. He stands out on the balcony and there's like an AUDIENCE of about a million people out there to see him.

But hey, I'm not the lifelong Catholic around here so sure, we'll call it a "congregation."

Now I know you're wondering, is Betty Brewster any relation to the Brewster House on Russian Road, the one you wanted to buy? Well I asked her and she said probably but they haven't been able to find the connection.

But that got her talking about her family. So I learned all about Miss Brewster's family history today, cause I can be a good listener when I have to. Anyway what else were we gonna talk about? It's not like I was gonna start lecturing her about my phylogenetic analysis.

So anyway I don't have all the names and dates, but here's the deal. They're all descended from one of the old missionary families, but not one of the famous ones you've heard of.

Betty had two cousins, a brother and sister. They're the ones that lived in the house where I'm staying now.

Their dad and his dad and probably his grandfather were all pastors at that little church. The brother was going to be next in line (no women pastors, it was one of those super old school New England kine churches where the guys wear buckle hats and burn witches). But I guess the brother was kind of a black sheep and it didn't work out. Of course I wanted to know more details about that, but she wouldn't tell me. Instead she goes:

Oh, my dear, Titus was bad enough in his ordinary way. But Priscilla…

Right, now I remember their names. Titus and Priscilla. So of course I ask, what about Priscilla, what did she do? (Remember the park ranger told me about a suicide in the house? I really wanted to ask Betty about it but if it was one of her cousins who killed themselves I'd really kick myself.)

Anyway, instead of dishing the dirt about Priscilla, Betty gets all evasive. She said, well, her life was hard, she buried her father and then her brother, and she sold the church to the Methodists.

Oh, and she told me this: Titus, the black-sheep brother, had a collection of "terrible" books that the father didn't know about, and wouldn't have approved of. So like I said the father died, then Titus was lost at sea, and Priscilla boxes up the books in a big crate.

Then what do you think she does with them?

Sell them? Donate them?

Nope, she gets a couple guys from town to drive the box down to the coast and dump it into the ocean.

I told her I thought it was a waste, but then Betty says,

"When a man believes there is no afterlife, he can be as wicked as he likes. She had to rid the house of all that wickedness, you see."

I didn't argue with her but come on, really? I mean look at me, I'm agnostic so I guess you could say I'm not convinced there's an after-life. And I don't even cheat on my husband even though no one who knew him would blame me.

On the other hand, maybe that is true for some people. Like you, Molly, no offense but if it wasn't for your Catholic guilt I know some people who'd turn up missing pretty quick.

Hey that reminds me, how's it going with Kayla?

Emma Nakamura, PhD

Professor of Biology

Mahina State University

mahina.edu

It is now quite lawful for a Catholic woman to avoid pregnancy by a resort to mathematics, though she is still forbidden to resort to physics and chemistry.

H. L. Mencken (1880-1956)

CHAPTER FIFTEEN

Emma Kano`opomaika`i Nakamura <em-makn@mahina.edu>

to Molly

No of course I didn't know, how would I? I was just joking, seriously. Was it an accident? How did it happen?

Emma Nakamura, PhD

Professor of Biology

Mahina State University

mahina.edu

What we know is really very, very little compared to what we still have to know.

Fabiola Gianotti (b. 1960)

CHAPTER SIXTEEN

Emma Kano`opomaika`i Nakamura <em-makn@mahina.edu>

to Molly

Of course I'd come down if I could but unless I can get a helicopter and a winch to lift me off the mountain, I'm stuck here for now.

Okay, obviously after what happened you can't go anywhere, especially now you don't have anyone to watch the baby.

I'll admit you were right about one thing. Looks like Kayla lied about being a non-smoker. I thought the smoke smell was your imagination but I guess not. If Kayla was sneaking a smoke that explains what she was doing in your shed.

(Unless someone planted a cigarette on her? Why would they do that though.)

I can't believe Kayla's dead. Are you sure it was really her? Did you see the body?

The police don't seriously think you had anything to do with it, do they? Maybe you need to call in Honey Akiona again. Is she still practicing in Mahina?

OK, believe it or not, I have something that'll take your mind off your problems, no disrespect to poor Kayla.

Pay attention, I'm not making this up.

I was trying to figure out once the road opens back up if there's a way to add a phone extension without spending a fortune to get someone up here (it's a fortune, believe me. I already called.) I was poking around and in the drawer of the night table way in the back was this old piece of paper stuck to the underside of the table top, if that makes sense.

I don't have the bandwidth to send a picture but here's what it says:

To the one who finds this:

I have taken a woman's life.

Pray for me. I fear I am beyond salvation.

The signature looks like it could be Priscilla Hodges. The sister half of the brother-sister pair that lived in this house. (And maybe the same person that hanged herself in the living room? I still wish the friendly park ranger hadn't told me about that.)

Was she talking about her own planned suicide? She wouldn't have been talking about her brother, because she said I've taken a woman's life, and anyway the brother was lost at sea.

So assuming the signature really is Priscilla Hodges. Who did Priscilla murder?

(Whoa, hold up, I mean WHOM did Priscilla murder, not who, geez, Molly, what are you, like the human spell-checker? Don't you have more important things to worry about, like feeding your baby and staying out of prison?)

I don't think Mahina PD will be interested, but if you think it's worth passing along to them go ahead. Or maybe just keep this email in case I'm murdered by ghosts.

Now the note could be a prank or a fake, and I'm not saying I think it's necessarily real. But on the other hand, why would anyone write this note then stuff it into the back of a drawer? Get that big brain to work, Molly, and figure it out. Also this will keep your mind off the whole Kayla thing.

Poor Kayla. I still can't believe she's dead. So weird.

Emma Nakamura, PhD

Professor of Biology

Mahina State University

mahina.edu

The imaginative child will become the imaginative man or woman most apt to create, to invent, and therefore to foster civilization.

L. Frank Baum (1856-1919)

CHAPTER SEVENTEEN

Emma Kano`opomaika`i Nakamura <emmakn@mahina.edu>

to Molly

The park ranger came by to check on me. Her name's Ellie. I thought about showing her that letter but I decided not to. She said the truck's still burning, and they didn't think it was worth the risk to try move it. So no one's going anywhere anytime soon. I asked Ellie if she checked on Betty Brewster cause Betty had told me she was camping nearby, and she said yeah, Betty's fine. So she didn't think it was weird that a 90 year old lady was camping by herself in the middle of the forest. It sounded like she knew Betty though, so maybe it's actually normal for her.

I did ask Ellie if she knew anything about the brother and sister that used to live here, and if it was either one of them who killed themselves here, but she said she didn't know much, cause the house reverted to the park way before her time, and she thinks the suicide story's just a rumor anyway.

THANKS FOR TELLING ME THAT NOW, HELPFUL PARK RANGER, YOU OWE ME LIKE FIVE NIGHTS OF SLEEP.

And then she tells me she's glad someone's finally making use of the house because in all the years she's been working up here no one's stayed here until me. So once again, very reassuring.

Hey, I know you're busy fighting off murder charges and everything, but you get a good internet connection. Can you check a few years back for suicides and murders on the island? I kind of want to know if any of this is true, the suicide and the supposed murder in the note I found.

I sure hope you made the right decision talking to Detective Medeiros. You know on detective shows you're always supposed to lawyer up and then they can't do anything to you. Did you get any idea from them how Kayla died? I think you have every right to ask, the whole thing happened right in your backyard.

Emma Nakamura, PhD

Professor of Biology

Mahina State University

mahina.edu

Do not undertake a scientific career in quest of fame or money. There are easier and better ways to reach them. Undertake it only if nothing else will satisfy you; for nothing else is probably what you will receive.

Cecilia Payne (1900-1979)

CHAPTER EIGHTEEN

Emma Kano`opomaika`i Nakamura <em-
makn@mahina.edu>

to Molly

Eh, thanks for the article on false confessions. I didn't even think of that angle. I guess it puts things in perspective. So maybe Priscilla Hodges, if that's who wrote it, had a guilty conscience about something. She might've even blamed herself for her brother's death or something.

I can't remember exactly what it said now. Did she literally write that she killed someone, or did she just say she was responsible for someone dying? I guess I should just haul my lazy carcass upstairs and find the note so I can check for myself.

. . .

Well now I can't find the stupid note. I thought I remembered putting it right on top of the nightstand but now it's not anywhere. Nettie comes into the house sometimes. Maybe she was doing that thing cats do where they go onto a shelf and bat everything onto the ground. Last time (and I do mean the last time cause I'm not going there again) I went to visit my college roommate, her cat got on the bathroom counter and swatted all my stuff off. I found my deodorant floating in the toilet.

Oh I know what though. I can drive down to the church and check the gravestones to see when the brother Titus Hodges died. Then if I ever find the stupid note I can compare the dates. Not sure what that'll tell us, but it's something to do anyway. Long as I'm stuck up here thanks to the stupid burning truck.

Now, let's go over what you told me:

So it's two in the afternoon, you hear banging on the back door, it's your gardener.

He says he found Kayla in the shed unconscious with a burning cigarette on the ground next to her. (OK, you were right about that, she was sneaking out to smoke. Chalk one up for your suspicious mind.) Now, was she lying on the ground, sitting up, what? See if you can find out. I can't think of any biting insect here that could kill someone, but maybe that's something to look into.

Does the gardener have a motive? I mean probably not, but you can look into it.

Unfortunately your gardener messed up the crime scene when he moved her. Probably left his footprints all over the place.

What happened to the cigarette? Did the police take it? If I could get down there I could bring it into my lab and see if there's anything in there that shouldn't be. Did someone poison it?

DID YOU? JK I know you didn't. Probably.

Did you look inside the shed after they took the body away? Better yet let Donnie do it, just in case there's something toxic in there. He's not breastfeeding.

Anyway, I still haven't found the Confession Letter. But: I was looking in the storage closet in the back for a trowel, and there was an inside door that's locked. I brought my bolt cutters (the ones I told Yoshi the Women's Studies department was handing out, he still believes it LOL) and cut off the lock and guess what was inside: No not skeletons or anything like that but a big pile of old books. Holy anticlimax! They look pretty old, though so they're probably worth something.

Emma Nakamura, PhD

Professor of Biology

Mahina State University

mahina.edu

In science it often happens that scientists say, 'You know that's a really good argument; my position is mistaken,' and then they would actually change their minds and you never hear that old view from them again. They really do it. It doesn't happen as often as it should, because scientists are human and change is sometimes painful. But it happens every day. I cannot recall the last time something like that happened in politics or religion.

Carl Sagan (1934-1996)

CHAPTER NINETEEN

Emma Kano`opomaika`I Nakamura <em-
makn@mahina.edu>

to Molly

Remember I told you I found a closet full of old books yesterday? I didn't like seeing them outside where they could get moldy so I brought them in. The books…are not what I'd expect at the home of a pastor, let's put it that way.

I actually went out to look for Betty's camp today. Is that weird? I wanted to ask her if she knew anything about the books. I mean, if those were the ones Priscilla Hodges didn't throw away, I sure would like to know how bad they'd have to be to get tossed over the cliff.

But before I could find Betty, the park ranger, Ellie, found me.

I still didn't want to tell her about the note, especially now that I don't have it anymore cause she'd think I was just making it up. So I cleverly told her I thought Betty had said something about a murder although I could've misheard, and I asked her had she heard anything about it. She said no, but also that this place was pretty isolated and lots of stuff could've happened without anyone noticing. Then she told me even now she's pretty much the closest thing to law enforcement we have up here, so that was really reassuring.

I was going to tell her about the books I found but I decided not to. Cause when you think about it, how is that even worth talking about? Oh, I was in a house and guess what I found BOOKS. And if I told her I cut off a padlock to get to them then that sounds like I'm confessing to B&E. So I just kept my mouth shut.

I also didn't tell her that someone has been calling me on the stupid landline EVERY SINGLE NIGHT (yes, it's true) and that I heard crying outside the house again and I'm pretty sure it wasn't the cat (I heard it again when Nettie was with me, and it definitely sounded like human sobbing). Cause that would make me sound crazy.

I did ask her if she'd seen Betty and whether Betty was okay (between you and me I don't think she's okay at all; I'm starting to suspect Betty's the one who's been calling me and hanging up, and I wouldn't be surprised if she'd let herself in and was poking around too. Who knows how long it's been since the keys were changed?) Anyway Ellie said she'd check on Betty later, and that was that.

So I didn't get to ask Betty about the books. I really want to talk to her now. This is what else I'd ask her, while I'm at it:

1. Who was the woman Priscilla (thinks she) killed?

2. Where is the body, if there is one?

3. What was the motive?

4. Did you take the note from my bedroom?

5. What's the deal with the books?

Emma Nakamura, PhD

Professor of Biology

Mahina State University

mahina.edu

This is an era of specialists, each of whom sees his own problem and is unaware of or intolerant of the larger frame into which it fits.

Rachel Carson (1907-1964)

CHAPTER TWENTY

Emma Kano`opomaika`i Nakamura <em-
makn@mahina.edu>

to Molly

I'm so sorry I had to miss Kayla's memorial service. I hope you told everyone I'm literally stuck behind a burning truck, otherwise I would've come down for sure.

So Donnie finally went to check out the shed? That's good, although he doesn't have your obsessive attention to detail. I'm not surprised he didn't find anything out of the ordinary.

But yeah, I get why you don't want to go in there. I mean, Kayla died in there and we still don't know what killed her.

So:

Gardening tools, an old bicycle, hedge clippings, pesticide, weed killer.

Seems like the only thing in there that might kill someone without leaving a mark would be the pesticides. Although you didn't see her close up, did you?

I'm sure it was some kind of accident or natural causes. Who would want to murder Kayla? But if getting a house alarm makes Donnie feel better you should probably let him do it.

Emma Nakamura, PhD

Professor of Biology

Mahina State University

mahina.edu

I hope when I get to Heaven I shall not find the women playing second fiddle.

Mary Watson Whitney (1847-1921)

CHAPTER TWENTY-ONE

Emma Kano`opomaika`i Nakamura <em-makn@mahina.edu>

to Molly

Did someone say house alarm? Turns out I'm the one that should have a house alarm, not that it would do any good way up here.

I had a break-in last night. I heard some noise last night and thought it was the cat, but someone got into the office. They pulled books off the bookshelf and messed up my desk (shut up I know when someone's moved my stuff). At least none of my things seemed to be missing when I checked.

So I called the police right away and told them what happened, and also that there's only two people who could've done it, Ellie the park ranger and Betty Brewster. I said it has to be one of them cause no one else can get up here cause of the road being blocked. So the lady I talked to said hold on let me check something, then she came back and said, they cleared the truck away early this morning.

So the bad news is the burglar could be anyone. Although somehow I don't think so. I know I didn't leave any doors unlocked. Whoever it was must've had a key.

The good news is the road's finally open again so I'm gonna make a trip to Mahina. I can bring my samples down to the lab, and let the house sit empty for a couple days. That way the burglars can come back and take whatever they want and leave me out of it.

Of course I'll come over to your place. Can you drink yet? Beer's supposed to be good for nursing mothers, right? I'll bring a couple six packs.

And before you ask, no I haven't told Yoshi about any of this. I'm gonna wait till he gets back from his retreat. That way we'll have something to talk about for a week or so at least.

Emma Nakamura, PhD

Professor of Biology

Mahina State University

mahina.edu

[Those] who have an excessive faith in their theories or in their ideas are not only poorly disposed to make discoveries, but they also make very poor observations.

Claude Bernard (1813-1878)

CHAPTER TWENTY-TWO

Emma Kano`opomaika`i Nakamura <em-
makn@mahina.edu>

to Molly

Well, that was a fun trip back down to civilization. I'm back up at the haunted house now. I can see why it would be hard to travel with the baby, wow, does she ever stop eating? When she gets older are you going to tell her that her nickname was "the termite?"

It's still kind of weird seeing you as a mom, but even weirder seeing Donnie as a doting dad, don't tell him I said that. And not that I wasn't happy to see him, but I couldn't exactly snoop around your shed looking for clues with him there, could I?

But the good thing is it sounds like Pat's on the case. (That's one super annoying thing about the bad connection here, I can't get Island Confidential. But I know he's snooping around!) It looks like Donnie's OK with Pat hanging around now. Progress, right?

So I had a visit from a nice police officer and I couldn't identify anything that was taken, except the letter. I showed him the picture I'd taken of it on my phone. But it wasn't my property to begin with, so he was all like, whatever. He said I should change the locks, but he couldn't really help me with anything else. I guess I need to talk to Ellie the park ranger about changing the locks since the house is part of the park now. Oh and I asked him if he'd heard about your case, you know, Kayla, and he said yeah but he can't tell me anything about it. He knows your friend Detective Medeiros though. He said everyone remembers that time when Medeiros got everyone on the island involved in rescuing some lady professor from the university who'd gotten herself stuck down in a lava tube on an abandoned property. Have you ever heard of such a thing LOL

Emma Nakamura, PhD

Professor of Biology

Mahina State University

mahina.edu

I believe there is no philosophical high-road in science, with epistemological signposts. No, we are in a jungle and find our way by trial and error, building our road behind us as we proceed.

Max Born (1882-1970)

CHAPTER TWENTY-THREE

Emma Kano`opomaika`i Nakamura <em-makn@mahina.edu>

to Molly

I had Ellie the park ranger come over today. The phone reception's so bad up here I had to text her. I told her I wanted to put in a request to have the locks changed. The cop that came to talk to me confirmed there were no signs of forced entry. So someone has a key.

Of course if Ellie's the one who broke in then that was a stupid thing to ask.

But why would Ellie want to poke around the house now, when she could've done it before I moved in? And what happened to the confession letter, did the same person break in earlier and steal it?

Ellie told me Betty packed up camp and went back down to her house on the coast, wherever that is. She didn't know.

Emma Nakamura, PhD

Professor of Biology

Mahina State University

mahina.edu

Many receive advice, only the wise profit from it.

Harper Lee (1926-2016)

CHAPTER TWENTY-FOUR

Emma Kano`opomaika`i Nakamura <em-makn@mahina.edu>

to Molly

I hope Donnie's feeling better now! Poor dude, I've never known him to get sick before.

I guess it was too much to try to clean out the shed on top of everything else. (What happened to the gardener anyway? Traumatized by finding a dead body? And what the heck was in that shed? You should ask him if he handled any pesticides. Some of those old-school ones are pretty scary.)

Things are getting back to normal here. As normal as things can be stuck in the mountains hours from civilization. The good news is, there's been no more break-ins, no more weird phone calls in the night, and no more spooky sobbing outside.

Funny how all that stuff stopped after Betty Brewster left. Hmm.

Of course it could have been Ellie doing that stuff and trying to pin it on Betty Brewster.

Or someone I haven't thought of.

Or, you know, ghosts.

Anyway, I'm taking advantage of the peace and quiet and getting a lot done.

Emma Nakamura, PhD

Professor of Biology

Mahina State University

mahina.edu

I do not like it, and I am sorry I ever had anything to do with it.

Erwin Schrödinger (1887-1961), speaking of quantum mechanics.

CHAPTER TWENTY-FIVE

Emma Kano`opomaika`i Nakamura <em-makn@mahina.edu>

to Molly

Okay, Molly, normally I'd tell you stop being paranoid but that is weird that your gardener came down with the same symptoms Donnie did. How's everyone now? How are you feeling? What about the baby?

Are you sure Donnie didn't get splashed by any liquids while he was cleaning out the shed? Were there any leaky old buckets or containers?

What else is in your backyard? You know what, I bet there's someone up at the Ag College who can help. They have people they send out to help with contamination whenever someone wants to use an old sugarcane field. (You wouldn't believe some of the nasty stuff sugar cultivation left in the soil. Dioxin, arsenic, you name it. I'm not saying all that stuff's in your backyard, but then again, I'm not saying it isn't.)

Ellie told me Betty Brewster's in the hospital. It might be a huge mistake but I think I'm going to visit her.

Emma Nakamura, PhD

Professor of Biology

Mahina State University

mahina.edu

When we try to pick out anything by itself, we find it is tied to everything else in the universe.

John Muir (1838-1914)

CHAPTER TWENTY-SIX

Emma Kano`opomaika`i Nakamura <em-makn@mahina.edu>

to Molly

Well I got some news. Want to hear it?

Yes, I know you do.

I visited Betty Brewster in the hospital up in Waimea.

(I didn't just barge in, I bought her some flowers from the gift shop. I'm not a monster, Molly.)

I was surprised to see how frail she looked all of a sudden. I mean sure she was old, but she was healthy enough to spend like a week camping in the forest. I know you couldn't do that. (On the other hand I think that's what put her in the hospital to begin with.)

So I felt kind of bad taking advantage of her weakened state to grill her but I got over it. I figured I'd start by asking her about the letter. And just as I'm about to open my mouth to say something I see this folded piece of paper on the little table next to her hospital bed…IT WAS THE CONFESSION LETTER! She had it with her right there! The letter that was in the house next to MY bed! Well, not my bed but the bed I was using, you know what I mean.

So I sat down on the chair and I reminded her who I was, and asked her how she was feeling.

So she goes,

"I am not long for this world."

That's how she talks. Kind of reminds me of you. So I said,

"Sorry to hear it. Can I do anything for you?"

She didn't say anything, so then I go to pick up the letter, and I'm about to ask her where she got it, and she grabs my wrist—holy old lady strength Molly, I don't think I have a grip like that even in the middle of paddling season!

Then I said,

"Did your cousin Priscilla really kill someone? Like she says in that letter?"

Betty just looked away and didn't answer me. So I figured I'd try something simpler. I asked,

"Did you break into the house?"

Still she doesn't answer.

"Was that you I heard crying outside?"

Still nothing. So then I say, look. I'm here, no judgement, I got nowhere to be and neither do you, why don't you tell me what's going on?

She kind of snorts and then she says,

"I cannot abide wickedness."

Well, Irony Alert ahead, Molly. Go pour yourself a glass of wine, fill up your water bottle, get Baby clamped on, and get ready.

Are you back? Okay. Here's the story.

So Titus, the black sheep brother with the naughty books (yes, those were his books I found, and by "naughty" I mean like Voltaire and Jefferson) was supposed to be next in line to be the pastor. You know, of the little church where their dad was pastor, and his dad before him, etcetera. But like I said, he was the black sheep, and not only was he rebellious and full of heretical ideas, turns out he got someone pregnant.

Well the dad set the girl up with the kid on another island and sent her checks to keep it all quiet. Now Priscilla's doing a slow burn this whole time. Why does she care? Cause first of all her brother and her used to be pretty close, and now they're not close anymore, cause he has this whole other family now. And it's not like she has anything going on. Second it's not like the family had a lot of money to begin with, and she sees the family fortune going to support her brother's little love nest.

Okay, so she spends years fuming about it, and then the final insult was that when the father died, he left Titus all the money and the church building and he left Priscilla the house and no cash. And all these years she'd been the one taking care of the father while Titus was out doing whatever, and I guess Priscilla decided she'd had enough. She made her brother a cup of laurel tea, which I guess is poisonous? You should look it up cause you have a good internet connection. I guess I could've stopped into a Starbucks and used their WiFi but I didn't think of it until I got back home.

And he drinks the tea and dies! So he wasn't lost at sea like it says on his gravestone!

Although in a way he was…

Priscilla packs her brother's body in a crate, and tells everyone he was lost at sea. Then she calls one of the parishioners who knows the family. She says Titus left behind all these awful books and she's already boxed them up and wants to get rid of them. So he helps her put the box in the truck and they drive down to the cliffs and dump the box into the ocean.

Meanwhile, this is back before the internet and cable news, so it takes the news a while to get to Titus's girlfriend. But eventually she does learn that Titus has died, and she comes to claim her part of the inheritance for her daughter. So she shows up at Priscilla's house.

Now, here she is, this woman who's stolen Priscilla's brother from her. So you guessed it, Priscilla invites her in for a cup of tea. This time she stages a suicide. She leaves a note saying she can't live without Titus.

So if you're keeping track: Father dies, leaves the house to Priscilla and everything else to Titus. Titus is "lost at sea" so Priscilla inherits his money. Baby mama and love child are out in the cold. Baby mama comes to ask for her share, Priscilla kills her and stages it as a suicide.

Now here's where it gets interesting: the mother of Titus's child was his first cousin. She and Priscilla looked enough alike that people thought the dead woman was Priscilla.

The cousin's name was Betty Brewster.

So I said, genius that I am, but you're Betty Brewster.

And she goes,

I'm Priscilla Hodges, you idiot. I thought you'd have figured it out by now.

So I asked, what about Betty Brewster's daughter?

She doesn't know any of it, Priscilla / Betty said. She was an innocent. I made certain she was raised by a respectable family. I've watched over her, in my own way. And she'll be well provided for when I'm gone.

Did you come into the house and take the letter? I asked. Did you move my glasses?

It was my house, she said. She sold it to the county, and somehow the park ended up taking over the care of it, but she still thought of it as hers.

So it was her poking around in the house, which was the house she grew up in. She just felt entitled to it, I guess, and no one had ever changed the locks.

She died later that night. The name on her chart was Betty Brewster. I think I'm the only one who knew who she really was. It's too bad. I wish I could've gotten to know her better. I've never been friends with a real murderer before. That I know of, anyway.

Emma Nakamura, PhD

Professor of Biology

Mahina State University

mahina.edu

...they are ill discoverers that think there is no land when they can see nothing but sea.

Francis Bacon (1561-1626)

CHAPTER TWENTY-SEVEN

Emma Kano`opomaika`i Nakamura <em-makn@mahina.edu>

to Molly

Eh Molly,

I can't believe I been to your house all those times and never noticed your hedge. I mean I did, but I didn't really think about what kind of plant it was. So I'll take the credit for inspiring you to find the information, but I guess you really need to thank the late Betty Brewster aka Priscilla Hodges for bringing it up.

Yes, an enclosed space like your shed full of laurel cuttings would definitely release enough cyanide gas to kill someone, especially someone tiny like Kayla. And yes, you did the right thing telling Detective Medeiros. Don't worry, when he checks with his experts he'll realize it was an accident, not a murder.

Good thing you found out now. You don't want a poisonous hedge with a baby in the house. A fence is probably going to be lower-maintenance anyway. Yeah, I know it means the guys with the chainsaws won't be going away anytime soon. But you know it's worth it in the long run.

It's weird to be packing up now. I'm almost used to being here. Ellie the park ranger stopped by to make sure everything was okay before I checked out. She's the one who told me "Betty Brewster" passed away. And maybe you saw this coming, I didn't.

Ellie's a rich woman now. "Betty Brewster" left everything to her. Ellie thought "Betty" was just someone who loved the outdoors as much as she did. She doesn't know "Betty" was her aunt. Not to mention the person who murdered Ellie's biological parents.

So there you go.

I can't wait to get back to Mahina. I know it's hard for you to get to a restaurant with all the baby paraphernalia, so how about I just come over and we can order in a pizza? Nettie and me should be back in Mahina by tomorrow afternoon. You're not allergic to cats, right?

Emma Nakamura, PhD

Professor of Biology

Mahina State University

mahina.edu

Happy is he who gets to know the reasons for things.

Virgil (70-19 BCE)

ABOUT THE AUTHOR

Like Emma Nakamura, Frankie Bow teaches at a public university. Unlike Emma, she is blessed with delightful students, sane colleagues, and a perfectly nice office chair. Follow Frankie's blog at frankiebow.com

Thank you for taking the time to read this book. If you enjoyed it, please consider telling your friends or posting a short review. Word of mouth is an author's best friend and much appreciated.

Mahalo, Frankie

Made in the USA
Monee, IL
11 November 2026

a6947a47-a838-4b32-b0a4-cd694c591912R03